# SHEEP FOR BEGINNERS

# SHEEP FOR BEGINNERS

## A DIP INTO THE WORLD OF WOOL

JOHN K. V. EUNSON

BLACK & WHITE PUBLISHING

First Published 2005
by Black & White Publishing Ltd
99 Giles Street, Edinburgh EH6 6BZ

Reprinted 2006

ISBN 13: 978 1 84502 040 8
ISBN 10: 1 84502 040 5



Illustrations are reproduced by kind permission of Michael Marshall.



A CIP catalogue record for this book is available from the British Library.

Printed and bound by Nørhaven Paperback A/S

# CONTENTS

# ACKNOWLEDGEMENTS

My thanks to Patricia Marshall
and all at Black & White Publishing.

To my parents and sister,
the Thomson family,
Vaila Angus
and
Margaret McShane's lovely daughter.

I

# HISTORY

## *And on the Third Day God Created Suffolks*

IT HAS BEEN calculated that there are over 500 mentions of sheep and lambs in the Old and New Testaments of the Holy Bible. This is over 500 times as many mentions as there are of cats in the Holy Bible – of which there are exactly none. Furthermore, these 500 references do not take into account the numerous times that the following are mentioned:

Flocks
Shepherds
Good shepherds
All right shepherds
Not so good shepherds
Wolves in sheep's clothing
Sheep wearing clothes in the first place

Sheep also played a large part in the conversion of Scotland to Christianity when, in 573 AD, Columba left Ireland with twelve followers to set up a monastery on the beautiful but remote island of Iona. It was to be a precarious and harsh existence of self-sufficiency. Without their loyal flock of sheep, it is probable that Columba and his monks would have starved or at least have had to eat more fish. This, in turn, could well have resulted in no Christianity in Scotland, no Reformation in the sixteenth century, no ladies wearing interesting hats on Sunday mornings and no turning of old churches into city-centre theme pubs. Furthermore, Christmas would not exist and would have to be replaced by a winter holiday comprising of nothing more than excessive eating and drinking and rampant retail consumerism.

However, all this talk of spirituality should not allow us to overlook how aggressive sheep can be. Anyone who has ever seen young rams sparring will have shuddered at the violence of such encounters. So ferocious can the encounters be that participants have been known to knock each other unconscious. This can act as something of a silver lining as it usually results in the dispute that caused the original collision being forgotten.

These warlike tendencies have, since ancient times, been utilised by man. It was only a last-minute failure of Greek craftsmen to carve realistic-looking wool accurately that saw a Wooden Horse rather than a Wooden Sheep end the Trojan War.

In Scottish history, sheep were present at Bannockburn in 1314, at Flodden Field in 1513 and at Culloden in 1746 – except for the couple of days in each of those years when two armies turned up and moved the sheep on so they could have a big battle.

Forty-six years after Culloden, Scotland had its very own Year of the Sheep in 1792. This was the year that the Cheviot was introduced into Caithness and the north of Scotland. Over the next sixty years, landowners throughout the Highlands and Islands replaced their traditional cattle with this new, more profitable sheep which had the advantage of being sold for both meat and wool. However, owning sheep required considerable land for them to graze on and the landowners' infamous solution to this problem was that their tenants would be moved out of their homes to make room. The Highland Clearances had begun.

The most infamous of the Clearances took place in Sutherland where thousands of people were moved off their land. The Countess of Sutherland and her factor Patrick Sellar became historical villains through their use of force and their callous burning of homes. Agricultural reform was sweeping through Europe in the eighteenth century but it was the insensitivity and brutality associated with the Highland Clearances that have neither been forgotten nor forgiven.

The irony of all this social and economic upheaval is that, by the end of the nineteenth century, sheep prices were in steep decline and the landowners started to replace them with deer. They had little option to do anything else as other types of land use required people and they had all gone.

So if you are travelling through some American backwoods, passing through Australian coastal towns or visiting a New Zealand north-country pass and you find yourself surrounded by Scottish place names, Highland surnames and reassuringly bad diets, then consider this – if it had not been for the introduction of a few Cheviots two hundred years ago, then all those Scottish ancestors who emigrated would have had to have thought of some other reason to leave.

As we can see, sheep and history are inextricably linked. The Chinese went even further by naming every twelfth year after sheep and, furthermore, they set out the characteristics of sheep as being lazy, timid, elegant, stylish, insecure, highly-strung, aesthetic and *Guardian* reading.

To conclude this section, listed below are a few famous historical events that happened in significant Years of the Sheep:

**143** — Romans complete the construction of the Antonine Wall to stop sheep sneaking in and eating their pasta

After struggling with the sudoku in today's

*Guardian*, Ramsay decided it was time for another nap.

**563** — Columba, his followers and some sheep arrive in Iona

**1307** — Edward I of England, the Hammer of the Scots, dies. Scotland breathes a sigh of relief. England is stuck with the Stone of Destiny for the next 700 years

**1559** — John Knox returns to Scotland from exile after successfully growing his beard to the correct Presbyterian length

**1643** — The Solemn League and Covenant is agreed between the Scots and English Parliaments – widely accepted as being not very funny

**1715** — First Jacobite Rebellion in Scotland – sheep remain neutral

**1919** — German fleet scuttled at Scapa Flow in Orkney despite the offer of lifelong supplies of fudge

**1979** — First referendum on Scottish devolution fails due to the Scottish public deciding that plans for a new parliament are just not expensive enough

Sheep were brought to the Americas by
the Spanish in the sixteenth century.
They brought multicoloured sheep
from Europe
which they swapped with the Navajo
and called them Noel.

2

# THE ARTS

*Much a Ewe About Tupping*

DESPITE THE FACT that it was the Lamb siblings, brother Charles and sister Mary, who wrote the classic *Tales from Shakespeare* for younger readers, there are actually very few references to sheep the works of the Bard.

To try to counter this cultural shortfall, a poll was conducted of shepherds in Britain and Ireland to find the top eight *Desert Island Sheep Discs*. The connection with the long-running radio programme being that the Bible and the *Complete Works of William Shakespeare* are the obligatory books given to castaways. The top eight were as follows:

1. 'Woolly Bully' by Sam the Sham and the Pharaohs

2. 'I Want You to Want Me' by Sheep Trick
3. 'Ewes Are the Sunshine of My Life' by Stevie Wonder
4. 'Wishing' by A Flock of Cheviots
5. 'Black Betty' by Ram Jam
6. 'Too Drunk to Suffolk' by Dead Kennedys
7. 'Mutton Compares to You' Sinead O'Connor
8. 'Lambing' by Bob Marley and the Wailers

A similar poll was conducted of Open University students who combined degrees in Livestock Management and English Literature to discover the Top Ten sheep-related novels of all time. They came up with the following list:

1. *A Room With A Ewe*
2. *Woollysses*
3. *Love in the Time of Scrapie*
4. *Three Sheep in a Boat*
5. *Sheep on the Road*
6. *Catch 22 Soay Sheep and You'll Be Lucky*
7. *1984 Was a Good Year for Wool Prices*
8. *Great Expectations for Wool Prices this Year*
9. *Remembrance of Wool Price Past*
10. *Tender Is the Lamb*

Scotland of course has a long literary tradition ranging from Robert Burns, Walter Scott and Robert Louis Stevenson to Lewis Grassic Gibbon, Muriel Spark and Irvine Welsh. And sheep have been ever present in this proud heritage.

Robert Burns was a farmer in Ayrshire who worked on his father's farm as a boy and took on another farm after his father's death. He wrote poems about sheep as well as the dogs, horses, mice and lice that he grew up with. Walter Scott spent a lot of his youth on his grandfather's farm in the Borders. Lewis Grassic Gibbon was born on an Aberdeenshire farm and set *Sunset Song*, the greatest Scottish novel of all time, in the countryside of the Mearns. James Hogg, the author of *Private Memoirs and Confessions of a Justified Sinner* was known as the Ettrick Shepherd because a) he was a shepherd and b) he came from Ettrick.

John Buchan wrote a novel called *The Island of Sheep*. This was an adventure story featuring Richard Hannay who was also the hero of *The Thirty-Nine Steps* which was turned into the

'Of course the three of us can fit in that boat!

C'mon, George, it'll be a laugh!'

classic 1935 Alfred Hitchcock film starring Robert Donat. In the pivotal moment, Hannay escapes from his kidnappers through the timely intervention of a friendly flock of sheep on a Highland bridge.

Finally, the irascible and now almost forgotten Hugh MacDiarmid came from sheep country in the Borders and featured sheep in his most famous poem *A Drunk Man Looks at the Thistle*. Here are several words he may have considered as possible rhymes with sheep:

**bleep** – CENSORED
**creep** – a cheerful Radiohead song
***L'Equipe*** – a French sports newspaper
**heap** – an old car
**jeep** – a suburban vehicle used for supermarket shopping
**neep** – a turnip or foolish person
**reap** – what reapers do, not to be feared
**Sweep** – a friend of Sooty
**weep** – an emotional response to watching Scotland play football or rugby

In the end, MacDiarmid rhymed 'sheep' with 'nether deep' (see *A Drunk Man Looks at the Thistle*, stanza 7).

In the world of art, sheep tend to have a peripheral role in the art galleries of the world. Appearing in the background of pastoral, rural and religious paintings, only Vincent Van Gogh of the Great Masters gave sheep their rightful place of prominence with *Shepherd with a Flock of Sheep* and *The Sheep-Shearers*.

Sheep however remain an integral part of Scottish art as highlighted by a recent survey of what are the most popular pictures in Scottish colouring books:

1. Nessie, the Loch Ness Monster – Green crayon
2. Greyfriars Bobby – Grey crayon
3. Pipers – Red and black crayons
4. Highland cows – Orange crayon
5. Castles – Grey crayon
6. Sheep – White crayon
7. Clansmen – Green crayon
8. Eagles – Gold crayon

9. Thistles – Purple crayon
10. Deer – Red crayon

Damien Hirst brought sheep into the shock of the new world of modern art with his exhibitions of sheep preserved in formaldehyde. Farmers were appalled by the exhibits, considering them a waste of a perfectly good fleece.

Most summer rock festivals take place on farmland. Sheep don't mind this disturbance but could really do without the Stereophonics droning on again.

Of all the arts, sheep are
keenest on the theatre.
They like to watch the dramatic
action on stage and appreciate the fact
that it is dry inside.
Their five favourite plays are:

*Hay Fever* – Noel Coward
*Sheepmalion* – George Bernard Shaw
*Cheviot Flock Circle* – Bertolt Brecht
*Lady Windermere's Ram* – Oscar Wilde
*Taming of the Ewe* – William Shakespeare

Sheep have good eyesight
except of course when the wool
is pulled
over their eyes. This is a
painful procedure
that requires an anaesthetic and only
became fashionable in
the 1960s when Beatlemania
and the number one hit
'She Loves Ewes'
became popular with young sheep.

3

# ECONOMICS

*One Nation Under a Hoof*

THERE ARE FIVE factors that farmers look for in finding the perfect sheep:

1. **A bold eye and an alert expression** Farmers will spend hours observing a flock to identify the sheep that best represent these two traits, the ideal being a sheep that can stare the farmer out whilst simultaneously raising an eyebrow.
2. **A long straight back** A long back means more wool and it is good for posture in later life. Young sheep are encouraged to lift properly from an early age.
3. **An easy walking action** Lambs are trained to walk in a relaxed, easygoing manner. They will be seen constantly

practising walking back and forth in fields up and down the country. Swaggering is considered unacceptable and could result in immediate death.

4. **A balanced mouth** There should be no overshoot of either upper or lower jaw and the teeth should be in good condition. Fully grown sheep will have thirty-two teeth and it is important they look after them – although flossing is still not very common.
5. **Large, even-sized testicles** The reason for this is straightforward. It is vital for a farmer that their rams are not lacking in self-confidence as rams have many ewes to impregnate and do not want to be rejected by ewes who are of the opinion that size really does matter. The testicles should also be firm to the touch although there is a wide discrepancy on what the optimum time for holding the testicles should be.

There are then three types of sheep in Britain:

*LONGWOOL* — These are debonair, bohemian, artistic but sometimes scruffy sheep. They became more visible in the 1960s as the permissive society led to shearing becoming optional. The main longwool ancestor is the **Leicester**.

*SHORTWOOL* — Traditional, responsible and respectable sheep, the shortwools are often perceived as being more virile than longwools. However, this theory is widely believed to be propagated by male shortwools to make them feel better about themselves. The main shortwool ancestor is the **South Down** which begat the present-day **Suffolk**.

*MOUNTAIN OR HILL* — Those with a fear of heights prefer the latter name. Smaller, hardier, self-sufficient sheep, they are often derided for lack of sophistication and unfriendliness to visitors. Mountain and hill sheep go where they want, eat what they want and basically don't care. The phrase 'like water off a duck's back' would have been apt for mountain sheep, except for the fact that wool absorbs water and they are not ducks.

**Blackface** sheep came to Scotland in the eighteenth century and have flourished due to their ability to prosper in inclement weather and the fact that they can live on heather as well as grass. There are four types of Blackface in Scotland. The **Perth** and **Lanark** both have heavy fleeces, whilst the **Lewis** and **Newton Stewart** are smaller.

The Blackface when crossed with the **Border Leicester** produces the crossbreed **Greyface**. If the Border Leicester is delicate, then the cross is a **Paleface**. If the Border Leicester is young, then the cross is a **Babyface**. And if the Border Leicester is an Italian football fan, then the cross is an **Interface**.

The **Blackface Suffolk** is one of the heaviest breeds of sheep in the world with rams frequently weighing 100 kg or more. Suffolks are known for being alert and muscular with good stamina and balance. A typical Suffolk ram is matched up with thirty-eight ewes per year, any more being considered promiscuous. Suffolks also live a long and active life with elderly Suffolks often playing bowls, helping schoolchildren cross roads and writing letters to newspapers.

The heaviest sheep of all time weighed 247 kg, just shy of a staggering 39 stone. The sheep in question was a ram named Stratford Whisper, so called because he did not have to make much noise to get noticed.

The gestation period of sheep is around the 148-day mark. The nearest contemporaries in the gestation table are goats at 150 days and pandas at 138 days, when they can be bothered.

The most popular names for ram and ewe prize lambs in 2001 were:

| **RAM** | **EWE** |
| --- | --- |
| Larry | Lucy |
| Barry | Charlotte |
| Gary | Pat |
| Harry | Sam |
| Alan | Pam |
| Adrian | Tulip |
| Matthew | Marigold |
| Mark | Lulu |
| Luke | Cilla |
| Ringo | Ringo |

Eunice wasn't the only one in the family to be surprised by the arrival of her baby brother.

**gimmer** – a one-year-old female sheep that has been clipped
**dimmer** – a switch for hiding dust in a room
**hogget** – a lamb between six months and a year old
**dog** – an animal that's not as clever as it thinks

Sheep skulls have a convex upper contour of the muzzle which leads to the appearance of a Roman nose. This explains why sheep are proud of their Latin heritage and motto:

*Veni, vedi, pavi*
(We came, we saw, we ate grass)

It is common for ewes to have twin lambs. The bond between twins can be so close that they can finish off each other's baas.

Being the black sheep of the family can actually prove to be an advantage in times of heavy snow, when it is much easier to be spotted. Being the black sheep, however, can also be a distinct disadvantage if trapped in a coal mine in the dark.

Due to their profession, rams have to be egalitarian in their duties. They may have their particular favourites – ewes that they want to spend additional time with – but it is in their job description that they must see to all ewes equally.

Rams' performances at tupping time are rated on the basis of three criteria:

the number of ewes impregnated
the number and quality of lambs born
the general satisfaction of the ewes

Farmers will then compare notes and produce a tup ten.

Rams are known to have a very high opinion of themselves. If they were made out of chocolate, they would eat themselves. Then they would feel very ill indeed as that would be an awful lot of chocolate.

Sheep rustling has a long and
ignoble history and,
according to the police,
it is now on the increase again.
Ruthless gangs will steal
prize rams or entire flocks
and send ransom notes to farmers
complete with tape recordings of
sheep bleating
and balls of freshly shorn wool
if their demands are not met.

# 4
# FOOD AND DRINK

## *The Mutton Report*

THE GREATEST SCOTSMAN of all time is generally considered to be Robert Burns – internationally renowned poet, songwriter, wit, womaniser, drinker, radical, conscience of New Year and the only Scotsman to have his birthday nationally honoured. All this and he was dead by the age of thirty-seven.

Furthermore, Robert Burns's birthday, the 25th of January, is celebrated by a Burns Supper where participants drink whisky, recite poetry and get through lashings of haggis and vegetables. The fact that the haggis is traditionally made from parts of a sheep's heart, liver and lungs, mixed with oatmeal, suet and salt and cooked in a bag made from a sheep's stomach is a grave disappointment to the many visitors expecting little, cuddly, hairy animals that run around the Highlands. Even a trip to a haggis plantation in Dumfries and Galloway to see the sheep stomachs harvested in January can leave tourists distinctly unimpressed.

If haggis does not appeal to your taste buds, there are numerous other sheep-based dishes for your delectation. A good introduction is mutton pie which goes as follows:

*Ingredients*

12 oz (345 grams) sheep
salt
pepper
6 tablespoons stock
8 oz (230 grams) flour
2 oz (50 grams) fat
water
milk
egg yolk (the non-white bit)

*Directions*

Make meat filling first. Chop meat. Switch oven on to 190 C. Make crust. Sieve flour and salt in bowl. Boil fat, water and milk together in saucepan. Pour hot concoction into middle of flour and knead until flat. Roll this pastry and divide into two-thirds and one-third. Cut two-thirds into two circles and press into pie tins.

Fill the tins with meat, adding stock, gravy or water. The remaining one-third pastry can be used to make lids. Make slit at top of each pie. Brush with water and egg yolk (the non-white bit). Cook for 35 minutes. Serve hot or cold.

*OTHER THINGS YOU NEED TO KNOW*

Where exactly do you buy mutton from these days?
Do you actually have a sieve?
Do not throw out the one-third of pastry as, if you do, there will be a lids crisis
Rolling pin – you probably don't have one of these either
What sort of shops sells pie tins?
Do you really, really want to do this?

If mutton pie does not appeal then you are probably unlikely to try the traditional dishes of Boiled Sheep's Tongues or Powsodie (Sheep's

Head Broth). The recipe for Boiled Tongues begins with the instruction 'wash the tongues well' which begs the question whether that should be done before or after the sheep dies and, more worryingly, implies that more than one tongue has to be eaten.

The introductory instructions for Powsodie are even more disturbing:

1. Soak the sheep's head overnight in lukewarm water
2. Remove the eyes
3. Split the head open and lay aside the brains
4. Clean head thoroughly

On the other hand:

11. Garnish with parsley

So that's alright then.

Of course it is completely wrong to decry mutton. Until the advent of lambs being imported from the southern hemisphere all year round, mutton was a staple of the British diet. The traditional definition being that meat from a sheep under two years of age was lamb and sheep over two years old should know better.

In the last fifty years, mutton has disappeared from the dinner table but fans of the more mature sheep have been fighting back. Prince Charles is one of the many enthusiasts who have campaigned for the return of mutton to haute cuisine. The key to good mutton is a solid layer of fat, which adds flavour to the meat, and hanging it up a fortnight before preparing it. Good mutton has a rich, dark, gamey taste. It is vital that the meat is cooked very slowly, for up to seven hours, and at a low heat. Traditionally, a caper sauce would be an accompaniment.

Leg of mutton is now the more favoured part of the sheep, replacing the shoulder, which used to be so popular that sixty British pubs still have the delightful name of Shoulder of Lamb.

In the traditional dish of 'reestit mutton', the meat is salted and hung up for a period of eight to ten weeks to cure and dry. This often takes place on the pulley next to the range or Rayburn so you should ensure that the mutton is not confused with socks or other articles of clothing.

Ewes' milk and ewes' cheese are a small but growing market. The most popular dairy sheep in Britain is the Friesland which produce 300–500 litres of milk per ewe per year. Roquefort is a famous French sheep's cheese.

A popular lamb dish is Irish stew. Ingredients are lamb, spuds, carrots, celery, onions, garlic, flour, stock, rosemary, thyme, parsley and, most importantly, Guinness. As any Irishman or woman will tell you, it is essential that you allow the stew to settle before serving.

In the Republic of Macedonia, it is not unheard of for shepherds to give alcohol to their flock. Motorists in downtown Skopje are advised to watch out for drunken shepherds and staggering sheep on a Saturday night. Sheep are, on the whole, good-natured with a drink in them. They become more talkative and enjoy a laugh and a joke with friends. Sheep like McEwan's lager or, with a meal, a nice glass of Lambrusco.

Sheep Dip is the name of a 40% pure malt Scotch whisky, known for its potency. Much like the other sheep dip, repeated exposure can lead to a sore head and feeling under the weather.

A social smoker, Alistair only has the odd cigarette when he goes out to the pub.

The Sheep Heid Inn is a famous and popular pub in Duddingston, a self-contained village not far from the city centre in Edinburgh. The pub dates from the sixteenth century and has a skittle alley. Ironically sheep are not allowed to play skittles in the pub as they are underage.

Sheep are, of course, vegetarians but, unlike most vegetarians, they do not eat fish.

Why mint sauce? Well, sometimes in life, peas are just not enough.

If sheep have a hangover,
they tend to
suffer in silence
and avoid anything
to do with the
hair of the dog.

According to Welsh folklore,
sheep are the only animals that
will eat the grass that
grows in fairy rings.
This, they say,
explains why
Welsh mutton
is the
best mutton in the world.
Other animals have
countered that they
do not eat the grass that grows in the
Welsh fairy rings for that
very reason –
they do not want to be eaten.

5

# INTERNATIONAL

## *Around the World I Searched for Ewes*

AS WITH GOOD wine and Scots, sheep travel well. They have made their home in far-flung and exotic parts of the globe, assimilating into their new environments, whilst maintaining their culture, traditions, language and the same number of legs.

Other than Britain and Ireland, the major countries where sheep are found are Argentina, Australia, China, India, Iran, New Zealand, South Africa and Turkey. However, as international citizens of the world, sheep are to be found on every continent. In the Antipodes, the main breed is the Merino, originally imported from Spain. Known for its fine wool, the Merino is adaptable and brash. It enjoys sport and Vegemite and tends to be called Dan.

Myfanwy decided it was time to pack up her belongings, leave hor

ıd seek her fortune in the big wide world.

In Britain, the Suffolk's long dominance has been challenged by European breeds, in particular the Charollais from France and especially the Texel from the Dutch island of the same name. The percentage of British ewes mated to Suffolks has fallen from over 50% to under 30% in the last thirty years. This change in sheep demographics has resulted in resentment from unemployed Suffolks towards their new European neighbours for taking their ewes and all the best land. These Suffolks have gone as far as setting up their own organisation, UTUP, to campaign for British-only mating with British ewes. To avoid the criticism of being a one-issue party, UTUP also supports Free Tibet and the reform of local council tax.

As a counterpoint to such international tensions, Sheep Olympics are now held every four years to foster good relations. The Olympics feature such events as running, walking slowly, swimming, jumping streams, chewing competitions and fencing (i.e. how many fences a sheep can get through in a given period of time). The Australian team usually wins the most medals, the Argentinians tend to fall over a lot and the British come a plucky third.

In New Zealand, as well as swimming with dolphins, you can swim with sheep. Otago Swimming Baths have Ewes' and Lambs' Beginners and Intermediate Classes on Tuesdays and Thursdays.

The Himalayas have more sheep breeds than any mountain range in the world. The Tibetan sheep is also the largest wild sheep in the world. The Marco Polo sheep is famous for its long spiralled horns and for having the same name as the Venetian explorer. One can only imagine Marco's reaction when discovering he was named after an Asian sheep.

Lanolin is an oil that is produced from sheep wool. It is used in the production of expensive cosmetics and creams.

The rain in Spain falls mainly on the plain. The rain in the West of Scotland just falls. No wonder that some sheep are just plain dour.

You can take a sheep to water
but you cannot make it
moisturise.

Many sheep are born as twins
or even triplets.
This perhaps explains why so many
people think
sheep all look the same.

6

# GEOGRAPHY

*Fear of a Woolly Planet*

IT IS THOUGHT that sheep were introduced to Scotland 4000 years ago. The oldest breed is believed to be the Soay from the St Kilda island group of the Outer Hebrides. The Soay can have different colours of fleece and both the ram and the ewe can have horns.

Where the Soay is unique amongst British breeds is that, due to its remote location, the Soay has been able to run wild for centuries. This has resulted in an interesting situation for farmers wishing to breed Soay on the mainland as, unlike other sheep, the Soay do not flock together. When approached by men or dogs, they simply scatter to all corners. Therefore, working with Soay requires a highly intensive ratio of one man and one dog

# THE BIG

*I heard she was told to bulk up for the role.*

Starring **Lauren**

# SHEEP

**Baaacall**

*Oh, really? I thought it was all to do with the clipping techniques.*

to one Soay sheep or hours of incessant running around until all parties are completely exhausted and agree to try again the following year.

As well as the Soay, the islands of St Kilda provide another breed of sheep in the form of the Boreray. Trouble is guaranteed whenever the Soay and the Boreray meet, especially if drink has been taken.

Sheep are found throughout the Western and Northern Isles of Scotland. In the most remote of these islands, it is the human population that has gone, leaving the sheep as masters of all they survey. The Hebridean breed comes from the Western Isles. Horned with all-black fleeces, they are quite good at rugby.

The North Ronaldsay breed hails from Orkney and has uniquely adapted to its Orcadian coastal habitat by incorporating seaweed into its diet and by bleating in a strange Welsh-sounding accent. The most populous of the island breeds is the Shetland sheep, with over 28,000 ewes. The Shetland is a small and hardy breed. They are known for fine meat, have multicoloured fleeces and tend to be called Joseph.

Shetland wool can come in many shades of white, grey, fawn, brown, moorit or black and is ideal for sheepskin rugs and for use in the world-famous Fair Isle knitting patterns. It is often claimed that the brightly coloured Fair Isle patterns are of Spanish origin. This came about because of Spanish Armada sailors fleeing north-wards after their defeat of 1588. They were shipwrecked off Fair Isle, one of the Shetland Islands lying about halfway between Orkney and mainland Shetland, and took up teaching the islanders Spanish patterns. Whilst it is true that the Spanish ship *El Gran Grifon* was wrecked in 1558, leaving 200 seamen stranded on the island, it is highly unlikely that this has any connection with the patterns. 'A recent study has found that Fair Isle designs are in the Scandinavian tradition and there is no evidence of any Spaniards remaining permanently on the isle,' concluded local historian Brian Carlos Garcia.

Sheep from the Northern Isles are transported to the mainland by overnight ferry. There is a bar, a restaurant and a mini-cinema on board the ferry to take their mind off any seasickness.

No household is complete without a beautiful sheepskin on the floor, bed or settee. Available in white, grey, brown or black, sheepskins are favoured over the alternative skins of cow, donkey, cat, dolphin or hedgehog.

If you are camping in the
Great British countryside, it is
not unusual
for you to be woken early
in the morning
by a local sheep.
If this happens, do not be alarmed
as incidences
of axe-wielding, bloodthirsty sheep
are actually quite rare.

If sheep did not exist
then sheepdogs would become lazy,
feckless animals,
aimlessly wandering the countryside
terrorising unsuspecting rabbits.
Many sheep already consider this
to be the case.

# 7
# ENVIRONMENT

## *White Sheep Can't Jump*

ONE OF THE characteristics of mountain and hill sheep is that they roam over a large area. This is partly due to the search for good land to graze on but also due to innate curiosity about what is on the other side. They've discovered that every fence has its weak point, all dykes have their loose stones and each stream has its crossing point. 'The grass is always greener on the other side' is a very apt saying for mountain and hill sheep because invariably it is.

This underestimated ingenuity of sheep to exploit the saying has recently been highlighted in West Yorkshire by the discovery that sheep have been successfully traversing cattle grids by using the rollover – a technique akin to a canoeist capsizing a canoe. As you can imagine, this can prove a tricky and delicate manoeuvre which requires pinpoint accuracy. A small miscalculation could result in a sheep becoming stuck in the grid which could end in the humiliation of the unfortunate one having to be rescued. On the outside chance, that could be achieved by other members of the flock if they happen to have some rope handy. Failing that, humiliation will be all the greater because a human will have to be called on.

Sheep are known to be able to predict the weather. This isn't that surprising really as they all watch the weekly forecast on Sunday morning's TV farming programme.

Of all the weather conditions that sheep have to put up with, snow is their least favourite. It's cold, there is a scarcity of food and they are not good at skiing. Severe cold can have serious repercussions. In the 'Big Chill' of 1983, there was one death and much talk of re-evaluating lives.

Some sheep do not like to graze in too much direct sunlight – so much so that they will try to break into neighbouring fields if they think they offer more shade. These sheep are considered to have crossed over to the dark side.

A sheep pen is the name for an enclosure for sheep. A traditional name in Scotland for such a pen is a cro. A sheep pen is also the name of a pen with a small plastic sheep stuck on top.

Sheep have hooves not feet. It is important that their hooves are kept in good condition to avoid diseases such as foot rot. Sheep with cloven hooves are considered to be bad as they have sold their souls to Satan. Dyslexic sheep with cloven hooves are considered to be festive as they have sold their soles to Santa

The practice of sheep-dipping has fallen out of favour. Farmers and crofters reported that they became ill after repeated contact with sheep dip and they blamed the chemicals in the liquid for this. It can only be imagined what trauma was done to the poor sheep who were immersed in the stuff and a helpline was set up for worried sheep to call in with their concerns.

Sheep are great fans of the game hide-and-seek. Expert players can hide for months at a time.

Even though the graveyard offered so many good hiding place

oseph had agreed to count all the way to a hundred before trying to find them.

There is little evidence of sheep and cats having much to do with one another.

The fox has been an enemy of the sheep for centuries. In recent years, however, there has been a dissipation of this conflict, with many foxes moving to the cities in search of a better standard of living and kebabs.

Through countless generations of having to contend with nettle stings, sheep have evolved and now know it's best to avoid nettles altogether.

Almost all sheep today are shorn by electric shears rather than old hand-held clippers. This has resulted in requesting a number one and tramlines becoming a far less risky option than it used to be. Outside the tupping season, rams being sheared will be offered something for the weekend.

Sheep are not considered to be ideal house pets. Having said that, neither are alligators or bears so don't let it put you off.

When it comes to predicting
the weather,
sheep have often
been used as an indicator.
There is the famous saying,
'If March comes in like a lamb,
it goes out like a lion;
if it comes in like a lion,
it goes out like a lamb.'
Which presumably is not
referring to having
your tails docked.
Another weather saying is,
'When sheep gather in a huddle,
tomorrow we will have a puddle.'
which is not only prescient
but it also rhymes.

If sheep could fly,
then what would be the point of clouds?

8

# MEDIA

*The Big Sheep*

IN 1999, THE BBC announced that, after twenty-six years, they were cancelling *One Man and His Dog*, a series of televised sheepdog trials, with heats in England, Scotland, Wales and Ireland, followed by a grand final. The announcement of the cancellation caused uproar. There were letters, protests, petitions and questions in parliament. It led to hundreds of thousands of people joining the Countryside Alliance and marching through the streets of London wearing waterproof clothing.

The official reason for the BBC's decision was that the ratings for *One Man and His Dog* had been in decline and that many of the people who were vociferously complaining had not watched the programme in years. In truth, the last few years of the programme had been dogged, if you'll pardon the expression, by controversy. There was constant gamesmanship by both the sheep and the sheepdogs, with intimidation on both sides, frequent feigning of injuries and, latterly, disputes over appearance fees and image rights. When agents on behalf of the sheep demanded that their clients be included in the programme title, it was to prove one demand too far and the programme was axed.

The 1970s provided another popular British television show that involved sheep with the children's series *Stories from Toytown*. The star of the programme was indisputably Larry the Lamb. For two years, the stories of S. G. Hulme Beaman were adapted into twenty-six ten-minute programmes which captivated children of all ages before the tragic dénouement where Larry is . . . k-k-k-killed and ea-ea-ea-eaten.

Sheep have also played a crucial role in the history of cinema. The most famous sheep movie of all time is *Silence of the Lambs*, the multi-Oscar winning 1991 thriller starring Jodie Foster and

Anthony Hopkins, followed closely by Ridley Scott's *Blade Runner* from 1982 which is based on Philip K. Dick's novel *Do Androids Dream of Electric Sheep?*

However, there are numerous other sheep references throughout Hollywood folklore. In *Citizen Kane* by Orson Welles, usually described as the greatest film of all time, when Charles Foster Kane whispers the word Rosebud on his death bed, could it be he's referring to a pet lamb rather than some unlikely old sledge?

In David Lean's *Lawrence of Arabia*, is that speck on the desert horizon really Omar Sharif or could it perhaps be a stray sheep looking for water?

In Martin Scorsese's *Taxi Driver*, Robert De Niro plays the increasingly deranged Travis Bickle. Is it his paranoia with sheep that leads to the famous 'Ewe talking to me?' speech?

In Quentin Tarantino's *Pulp Fiction*, there has been considerable speculation about what is in the briefcase that Samuel L. Jackson and John Travolta are carrying when Tim Roth holds up the diner. Could it be possibly be a beautiful golden woolly jumper?

Travis just couldn't understand why he was

finding it so difficult to pick up fares.

Long before the glitzy eye-candy of today's *Emmerdale*, sheep provided the backbone for the first eighteen years of its earlier incarnation, *Emmerdale Farm*. Indisputably more charismatic than either Jack Sugden or Alan Turner, the *Emmerdale* sheep were only ever out-acted by Seth Armstrong's moustache.

Farmers equate their sheep to Geri Halliwell. As in, it is preferable when both are carrying a bit of weight.

The most famous sheep superhero has a deep voice, is called Helen and spreads happiness around the world by exercise and regular walking. Yes, it's Helen Sheep Hero.

Sheep are known for having a wry sense of humour and an appreciation of irony. Their least favourite comedians are Jerry Ewis and Bernard Lambing.

When a poll was conducted of sheep to
discover what their
favourite album of all-time was,
a majority of castrated males
unsurprisingly
plumped for the Sex Pistols'
*Never Mind the Bollocks.*

# 9
# SOCIETY

*He Ain't Heavy, He's a Suffolk*

CALVINIST PREACHERS SUCH as John Knox brought the Reformation to Scotland in the 1540s. Within two decades, Scotland's political and religious institutions had been changed from Catholic to Presbyterian Protestantism. For the next 400 years, the Church of Scotland was to be the major influence on Scottish society. This included the protection of Sunday as a godly day of rest and worship with work and recreation strictly not allowed. These strictures also applied to farmers and shepherds who, in turn, gave their flocks the day off to sleep in, visit friends and take part in gambolling competitions.

Even today, in these more secular times, on a Sunday, you can still see groups of sheep meeting at a fence or a dyke, chewing the cud and discussing local news and the week's events. Gambolling, however, has become a major problem with the innocent frolics of yesteryear having been replaced by serious competition with considerable money being wagered at illegal gambolling dens.

Although sheep have always made the most of Sundays, there is little evidence of other festivals being celebrated. The period over Christmas and New Year is an especially melancholy and mournful one. The bad weather is made worse by memories of family and friends recently departed. Easter is of course stressful with lambing just around the corner. Halloween has too many strange people wandering around in the dark. Bonfire Night has both noisy fireworks and too many associations with Foxes and Bank Holidays see just too much traffic on the road for comfort.

It has been accepted in this era of Prunes Appreciation Day that it was time for respect to be shown and World Sheep Day was introduced. The 19th of July was the chosen date as the weather would be favourable and the lambs boisterous. People are encouraged to wear a woolly garment for the day and to hug a sheep or proffer mints. There is much work to be done in promoting World Sheep Day as, so far, only Albania, Andorra and Armenia have given official recognition in a curiously chronological collection of countries.

There is one football club in the north-east of Scotland that is inextricably associated with sheep. That club is, of course . . . Dundee Ewenited.

As every child knows that, if you cross a sheep with a kangaroo, you get a woolly jumper. However, if you cross a sheep with devout Christian and Olympic champion Jonathan Edwards, you get a holey woolly jumper.

Sheep have often been blamed for the theft of flowers from gardens. Florists have taken advantage of this stereotyping by leaving traces of wool at the scene of the crime.

What with castration, tail-docking and ear-marking, sheep have an understandable aversion to sharp instruments. Tattoos and body-piercings are frowned upon.

DUNDEE

Phones4ewe

‘One Eamonn Baaaaaanon, there’s only one Eamonn Baaaaaano

ne Eamonn Baaaaaanon, there's only one Eamonn Baaaaaanon!'

Sheep do not fetch sticks. They disdainfully laugh in the face of anyone that does and call them 'Sticky'.

The love that dare not bleat its name has cast a shadow over the good name of both sheep and shepherds for many years. As far as we are concerned, such speculation is mere rumour and tittle-tattle and Wellington boots are nothing more than comfortable waterproof footwear.

Sheep live in a matriarchal society with the fathers absent. A recent pressure group has been set up called Rams for Justice which encourages rams to wear fancy dress in public.

Sheep are great believers in discipline. Ewes will keep lambs in line by a swift thrust of the head. This is known as a dunt. The ewe is the dunter. The lamb is the duntee. And the past tense of dunt is dunnit.

Although ewes believe in discipline, they can also be indulgent to their cocky offspring. With the chances of a long and peaceful life for lambs somewhat limited, they might as well allow them to make the most of it.

Personal grooming is of vital importance to sheep. Any laxness in the toilette department, for example, soon shows up on white wool.

Sheep-ish
is defined as meaning abashed or
embarrassed.
Sheep-pish
is something different but with
the same result.

When standing on the edge of a cliff
and faced with a flock of sheep,
the one thing you should not say is,
'Hey, sheep!'
Sheep are not stupid and will suspect
that the hay
is just behind you.

IO

# COMMUNICATIONS

## *Do Vampires Dream of Immortal Sheep?*

FROM THE AGE of childhood, we are taught that the cure for insomnia is the counting of sheep. The theory being that the bucolic langour of an infinite flock streaming past in a blur of wool would send even the most restless soul to his slumbers. The problem with this theory is when it comes to those who actually work on a regular basis with sheep. If, for example, you have a flock of forty-two sheep, what do you do when you get to number forty-two and you are still awake? Or, even worse, what happens if you end up counting your flock and find that there are some missing? The worry of where they might have gone is bound to keep you up all night.

Shona hoped the other lambs would be so busy admiring her nev

choolbag they wouldn't notice that she'd forgotten her blazer.

There is, however, more to counting sheep than just trying to fall asleep. Throughout Britain, sheep-counting has its own traditional language which has remained to this day, despite the decline of the local dialects from which they came. There are linguistic variations of the words in different parts of the country but a general consensus goes something like this:

| | |
|---|---|
| one | yan |
| two | tan |
| three | tether |
| four | mether |
| five | pimp or mumph |
| six | tayter or yan a mumph |
| seven | layter or tan a mumph |
| eight | overa or tether a mumph |
| nine | dovera or mether a mumph |
| ten | dick |
| fifteen | bumfit |
| twenty | jiggit |

The University of Manitoba has produced many years of fascinating research into sheep language. There are hundreds of hours of recordings of the sounds sheep make when they are shown images of various objects. By 2001, the university had pinpointed sixty-three different sheep phrases based on tone, pitch and volume. For example, there is a baa that expresses fear and uncertainty. There is a baa which means 'Hello!' or 'Where are you?'. There is another baa that is used to suggest curiosity, interest or arousal. And there is a very long and relaxed baa which means 'Good grass, man!'

Sheep do not attend school unless there is the necessary access available for grazing on playing fields.

Sheep do have the facility to learn and most sheep would pass media studies if given the chance.

Rams are not known for their witty repartee and sensitive approach. With up to forty ewes for each ram to get round per breeding season, there isn't really the time.

Do not ask a sheep a direct question. Unless you know the sheep particularly well, you will be singularly ignored for such an abrupt and impolite approach.

Sheep do not play the National Lotto.
The odds are too high and
they object to this
additional indirect form of taxation.
If sheep won a lottery however,
they would buy a big pile in
Hertfordshire –
not for the prestige and
the market value of
the property,
just for the acres of
good grazing land
that go with it.

Sheep are not very good at juggling.
This has not proved to be
a hindrance in their
day-to-day lives.

11

# POLITICS

## *Ewes of the World*

TRAVELLING BY CAR through the Highlands and Islands of Scotland can be a precarious business. Single-track winding roads, blind corners, poor visibility and, of course, sheep on the open road are all factors that have to be considered in any journey.

So serious did the Department of Transport take this issue that a study was made on the probability of sheep crossing the road as a vehicle approaches. After three years of research, the findings were inconclusive. When asked:

38% of sheep said they never crossed the road
36% of sheep said they sometimes crossed the road
22% of sheep said that crossing the road was a great adrenaline rush

Christina was getting really worried – if the bus didn't come alon

soon, she wouldn't make it to her euphonium lesson in time.

4% of sheep unfortunately failed to complete the study

2001 was a dreadful year for British sheep. The reason for this was something worse than traffic accidents. The foot-and-mouth epidemic began in February and it was officially the worst outbreak anywhere in the world. Over six million sheep and cattle were killed and 9000 farms were affected as the crisis became forever associated with burning pyres of dead carcasses.

Foot-and-mouth is not a fatal disease for sheep – a full recovery will take place naturally – and it is not directly transferable to humans. It was, therefore, somewhat disappointing to sheep that the government chose the culling rather than the vaccination option. Not surprisingly, visits to the vet with minor ailments have subsequently been drastically reduced.

The most famous female country-and-western singer of all time is, of course, Dolly Parton. An illustrious career as singer and songwriter that includes such classics as 'Jolene', 'The Coat of Many Colours', 'I Will Always Love You' and 'Nine to Five' has made her a popular and respected international star. It was, however, the singer's considerable cleavage that inspired the cerebral and renowned Roslin Institute to christen the world's first cloned sheep 'Dolly' as the cloning process had involved tissue from a mammary gland.

Dolly was created in July 1996 at Roslin, near Edinburgh, which is near the world-famous Rosslyn Chapel, one of a select number of locations featuring in the Holy-Grail-Is-Not-Here-Either Grand Tour. Dolly was a Finn Dorset breed, with a curious Bournemouth–Helsinki ancestry. Dolly gave birth to a lamb called Bonnie, named after another country singer Bonnie Raitt, in 1998 and she produced four lambs in total. By 2002 and at the age of five, Dolly was showing signs of arthritis and premature ageing. She was finally put down in February 2003 after suffering weeks of extreme déjà vu.

A ewe needs a ram like a female fish needs a male fish. Feminism has yet to reach the world of sheep to be honest.

It is often claimed that it is mainly sheep who vote Labour in Scotland. This is, of course, deeply unfair to the majority of sheep who are far more liberal in their views.

The gate to happiness for a sheep
is not a gate.
Sheep do not like gates.

Of all the team sports, sheep are
especially fond of rugby.
Australia, New Zealand, England, Wales,
Ireland and France
are all top rugby nations and,
coincidentally,
they all have large sheep populations.
The most popular rugby team for
sheep to support are the Barabarians –
who are
affectionately known as
the Baa-Baas.

# EPILOGUE

SINCE ANCIENT TIMES, sheep have proved to be vital components in the society that we live in. They were at the birth of the great religions of Judaism, Christianity and Islam. They were the driving force behind the economic growth of the great European powers from medieval times onwards. They were the backbone of colonial expansion to Australasia and the Americas.

Sheep have inspired artists, authors, poets and musicians throughout the centuries, culminating in the 1970s with Shari Lewis and her hand puppet Lamb Chop whose unremitting cuteness turned thousands of impressionable children into vegetarians.

Sheep have provided delicious cuisine and fantastic knitwear to millions. Where would we be if there were no lamb samosas or Pringle knitwear in the world? Cuddlier than cats, hornier than horses (as in horses do not have horns), prettier than pigs and more gregarious than goats, sheep are the daddies of farm animals.

As those great Scottish–Australian rockers AC/DC oh-so-succinctly put it:

FOR THOSE ABOUT TO FLOCK,
WE SALUTE EWES

As the Chinese proverb says,
'The sheep has no choice when in the jaws of a wolf.'
Unless, of course, the wolf is a friend and is just having a bit of fun.